EXTREME DISGUISES

BLACKBIRCH PRESS

An imprint of Thomson Gale, a part of The Thomson Corporation

Detroit • New York • San Francisco • San Diego • New Haven, Conn. • Waterville, Maine • London • Munich

THOMSON
GALE

For more information, contact
Blackbirch Press
27500 Drake Rd.
Farmington Hills, MI 48331-3535
Or you can visit our Internet site at http://www.gale.com

Photo credits: Cover: top left, bottom right © Photos.com; top center, top right, middle right, bottom left Corel Corporation; middle left © Digital Stock; interior: all pages © Discovery Communications, Inc. except for pages 4, 10 (top), 12, 16, 28 Corel Corporation; page 8 © Chris Mellor/Lonely Planet Images; page 10 (bottom) © Comstock; page 20 © Robert Yin/CORBIS; page 24 © David A. Northcott/CORBIS; page 36 © Photos.com; page 40 © Digital Stock

LIBRARY OF CONGRESS CATALOGING-IN-PUBLICATION DATA

Disguises / John Woodward, book editor.
p. cm. — (Planet's most extreme)
Includes bibliographical references and index.
ISBN 1-4103-0393-4 (hardcover : alk. paper) — ISBN 1-4103-0435-3 (paper cover : alk. paper)
1. Camouflage (Biology)—Juvenile literature. I. Woodward, John, 1958– II. Series.

QL767.D57 2005
591.47'2—dc22

2004019634

Printed in the United States of America
10 9 8 7 6 5 4 3 2 1

Get ready to find out what it takes to be a master of disguise and see if we can compete with nature's real-life transformers. We're counting down the top ten most extreme disguises in the animal kingdom and comparing them with our own attempts at cunning camouflage. You just won't believe your eyes when you see what happens when disguise is taken to The Most Extreme.

The Snow Leopard

Our search for the most extreme disguises on the planet begins in one of the most extreme environments. Hiding high in the Himalayas is number ten in the countdown: the snow leopard.

The snow leopard loves the high life. It can be found at altitudes above 16,000 feet, making it the highest living cat on earth. The snow leopard couldn't survive without some really cunning camouflage. That's why it spends most of its time posing as a big furry rock.

The snow leopard's spots help it to blend in perfectly with the rocky environment of the Himalayas.

On the trails of the Himalayas, leopard skin is the most fashionable form of disguise. On the catwalks of Paris, however, fake furs are all about getting noticed. People have tailored all sorts of animal camouflage into fashion statements.

How does a leopard manage to keep such a low profile? The answer lies in its incredible fur. Take away those spots and you're left with the easily recognizable shape of a hunting cat. But if your fur is exactly the same color as your environment, you blend in perfectly.

The trouble is, the snow leopard's environment isn't all one color. That's why the leopard has its spots. They disrupt the outline of the leopard, making it blend into the background.

Except for a weekly trip to the grocery store, Tom Leopard lives just like a real leopard, right down to his spot tattoos.

Spots make one man really stand out from the crowd, however. Tom Leopard is completely covered in leopard spot tattoos. For the past fifteen years, Tom has been the only leopard living wild in Scotland. This isn't part of a disguise, however. It's a lifestyle choice. Despite his lack of fur, he enjoys a rugged, reclusive lifestyle, just like a real leopard. A weekly hunt at the store provides him with his favorite meals and all the supplies any human leopard would ever need.

Unfortunately, there are very few stores in the Himalayas, so this leopard will never find cans of baked bharal. To catch these extremely athletic mountain sheep, the snow leopard has to rely on its excellent disguise and feline cunning. All it takes is one unsuspecting bharal to step into pouncing range.

Even though the snow leopard is the top predator in the Himalayan death zone, it must be careful. A fresh kill attracts unwanted attention to a creature that loves being invisible. The splash of red and the smell of blood destroy the leopard's careful camouflage, so it eats quickly. The snow leopard's extreme disguise means that while it's seldom seen, it's always spotted.

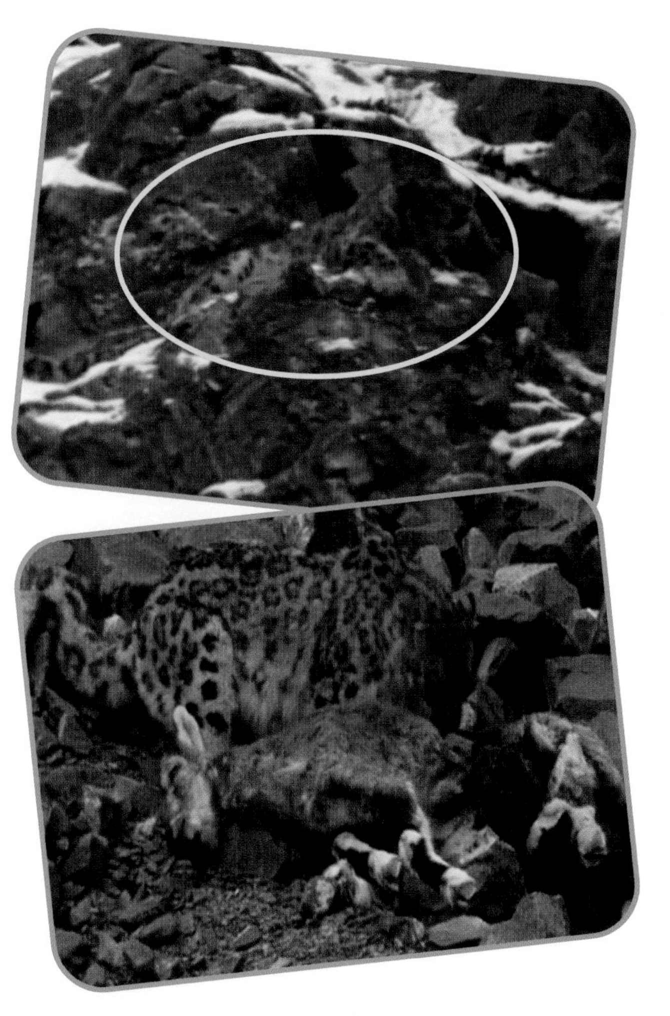

To catch a bharal for dinner, the snow leopard hides among the rocks (top) and waits for the chance to pounce (bottom).

The Lyrebird

Disguising the human voice can be tricky. Even our most famous ventriloquists can come across a little wooden. But number nine in the countdown is no dummy. It's the best impersonator in the Australian rain forest. Meet the superb lyrebird.

Male lyrebirds (above) can imitate over twenty different birdcalls to attract mates. They can even fool a real kookaburra (inset).

Nothing is better at disguising its voice than a male lyrebird who's looking for a mate. He builds himself a stage and then starts his crazy karaoke. His favorite hits include adaptations of over twenty different birdcalls. He's so good that even the real kookaburras are fooled.

The lyrebird's repertoire also includes some nonbird sounds, such as a camera with a motor drive, a car alarm, and even the sound of a chainsaw! The reason for this incredible mimicry is that the male with the most complex love song will attract the most females.

Male lyrebirds even imitate the sound of a chainsaw or a camera to attract as many females as possible.

A master of vocal effects, Michael Winslow claims he can imitate over 10,000 sounds. That's more than the lyrebird!

Far from the forests of Australia, there is one man with a voice as versatile as a lyrebird—a man who boasts a collection of over 10,000 vocalizations. Winslow and the lyrebird both practiced for years to perfect their sounds.

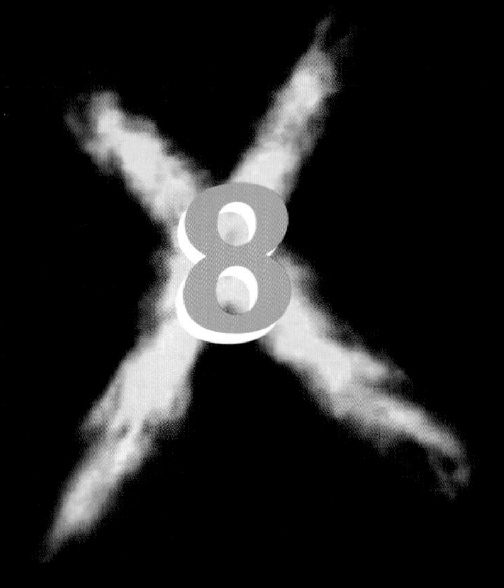

The Arctic Fox

In the Arctic Circle, a good disguise is the difference between life and death. Number eight in the countdown is the arctic fox. Its cunning costume is a real lifesaver, keeping it toasty warm even when temperatures fall to 110 degrees below zero!

An arctic fox eats the leftovers (inset) from a polar bear's (above) kill, staying far enough away to avoid being the bear's dessert!

The real key to staying alive out here, however, is the color of its coat. The white camouflage increases the fox's chances of a successful hunt and decreases the chances of it becoming the hunted.

The fox likes to stay close enough to a polar bear that it can clean up the leftovers from the polar bear's kill. It also has to stay far enough away to keep out of trouble, however. A polar bear would kill and eat an arctic fox. In this frozen land, being able to disguise yourself as a snowdrift is a huge advantage when following in the footsteps of a polar bear. Most of us, however, would need more than just camouflage if we met a hungry bear of any kind!

Specially designed for the movies, this bear head uses remote control and a human actor to create a realistic bear.

There is a very unusual bear in California. It was constructed by Robert Devine and his team from Anatomorphex Special Effects, which disguises humans for Hollywood movies and TV commercials. It takes more than just foam rubber, latex, fiberglass, and fur to make a good disguise, as Devine explains:

> *There is a live suit performer and five radio-operated performers to get all of the components running. Building these bears is like magic. It's a really wild way of making a living, if you ask me.*

Back in the Arctic, there has been more than one transformation. The snow has melted, and the fox has a new family and a new fur coat. The heavy white winter coat disappeared when the fox molted, or shed its fur. As summer approaches and there are more hours of daylight, hormones make the fox molt and grow a new, different coat.

Now the whole family wears lightweight summer coats. It's no coincidence that their new colors are great camouflage. It's this coat of many colors that lets the crafty arctic fox sneak into number eight in the countdown.

A fox family enjoys their new lightweight coats. By summer, arctic foxes have shed their heavy winter fur.

7 The Sea Snake

Sometimes when you're looking for contenders in our count-down of extreme disguises, you have to get down and dirty. On the Pacific Island of Vanuatu, disguise is a messy business. Men glorify the power of the animal that's number seven in the countdown. They perform a dance that tells how all the fishes in the sea were once saved from a very hungry shark, thanks to our next contender: the sea snake.

The black and white stripes of the sea snake are certainly distinctive and can also play clever tricks on the eyes. By weaving through the water, the high-contrast tail generates an optical illusion. It creates the impression that the snake is moving in the opposite direction of where it's actually heading. That means a hungry predator may be tempted to attack the snake from behind. This gives the snake the option of fleeing with only damage to its tail, or turning and fighting back.

Men in Vanuatu perform a sea snake dance (top). A banded snake eel mimics the movement of a sea snake (bottom).

There are very few animals in the sea brave enough to take on a sea snake. It is related to the cobra and has teeth that can inject powerful nerve toxins. There is one fish that has capitalized on the snake's bold designs and natural warning system. Meet the banded snake eel—a fish disguised as a snake! By looking and moving like a snake, this eel uses false advertising to keep predators at bay. It's a remarkably effective disguise, especially for an eel whose mark is worse than its bite.

Sea snakes aren't the only ones left seeing double. We've all been fooled by famous mimics. Although Adolf Hitler was all over Europe during World War II, some historians claim that it wasn't always him. Many world leaders employed body doubles as a decoy.

Even today some people make a living as look-alikes. Brent Mendenhall ran a construction business in Missouri until one day when his life changed. Brent now earns a living by impersonating President George W. Bush!

Some historians believe that Adolf Hitler used a body double for security reasons during World War II.

This man makes a living by working as a George W. Bush look-alike, making appearances across the country.

Mendenhall explains:

> *They say that everybody has a face double or twin. It's just that mine is a politician who some people say is the most powerful man on Earth. The first time I saw a picture of George W. Bush was in* Time *magazine in May of 1999. I was very surprised when I saw how much alike we looked. I almost thought it was a picture of myself in the magazine. The best part of the job is getting to meet a cross section of Americans all across the country as I work.*

6 The Spider Crab

Today, every soldier knows how to use camouflage to hide from enemy eyes. But our next contender has been practicing this branch of disguise for longer than any army! Down on the seafloor you can find a veteran campaigner of camouflage. Hiding away at number six in the countdown is the spider crab. This crustacean keeps its movements classified as it hides under its homemade canopy. It's no wonder that it is commonly called the "camouflage crab."

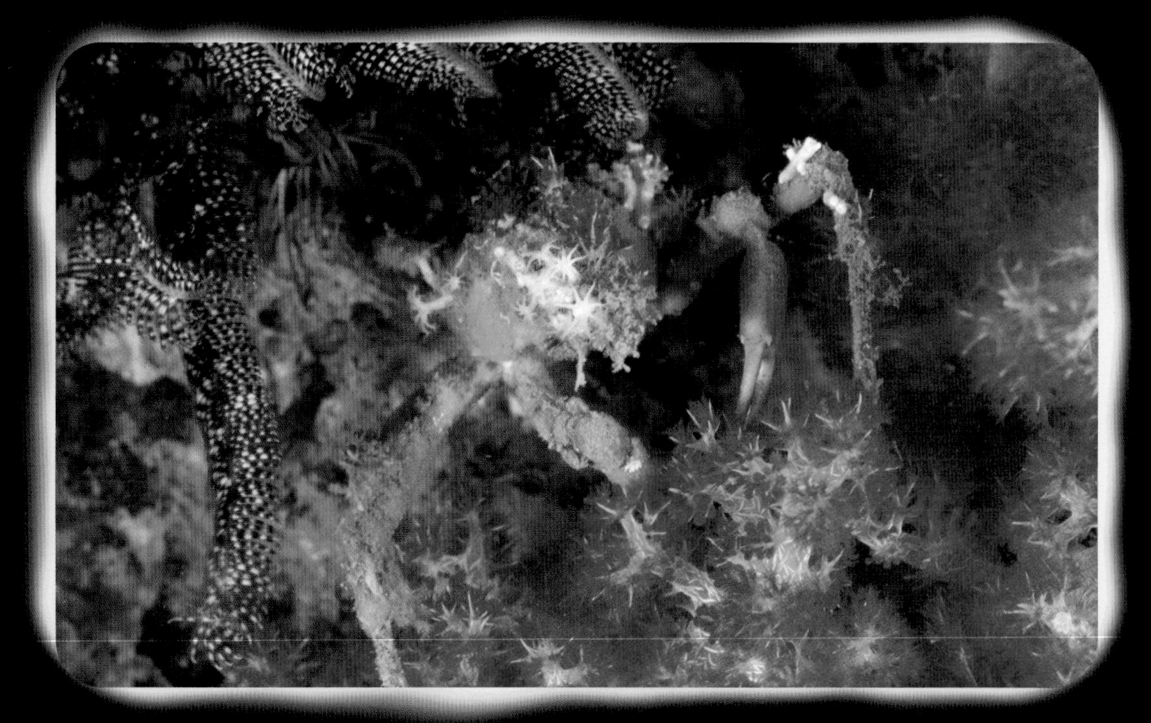

A master of disguise, the spider crab (above) creates its own leafy canopy by sticking plant pieces to its body (inset).

Armed with pincers, it trims pieces of plants and fastens them to tiny hairs on its legs and back. It works like Velcro. The chewed ends of the seaweed stick onto the hooklike body hairs. What makes the crab really extreme is that it changes its disguise when it changes environments.

Before World War I, military uniforms were brightly colored (above). Today, soldiers use camouflage to hide from the enemy (inset).

Modern soldiers select their camouflage to fit in with specific landscapes, just like the crab. These designs disguise both people and their military vehicles. What's really surprising is how long it took for military camouflage to really take off.

Before World War I, military uniforms were bright and colorful. Back then, camouflage uniforms weren't necessary. After all, the weapons of the time were wildly inaccurate. Why wear camouflage and try to hide if your enemy can't hit anything unless it's standing right in front of him? On a modern battlefield, however, weapons are very accurate and can strike from long distance. Wearing brightly colored uniforms would be like wearing a brightly painted bullseye!

Even today's most elaborately disguised soldier is no match for this crab. After all, what commando can eat his own camouflage when his rations run low?

A heavily disguised soldier can't eat his own camouflage the way a spider crab can.

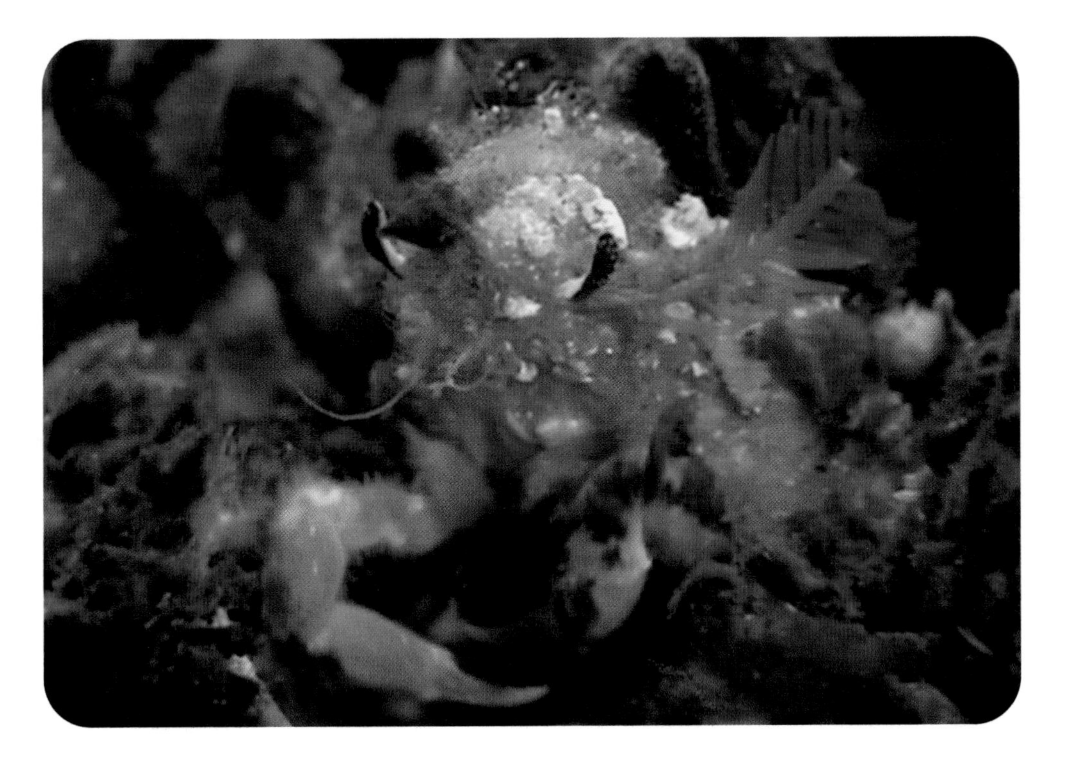

5

The Hognose Snake

Our countdown's heading out west on the trail of the most wanted disguises on the planet. That's because the next contender uses a gimmick that comes straight from a Hollywood western.

Sliding into number five in the countdown is the hognose snake. Common throughout America, these little reptiles aren't venomous, so they've had to come up with an extreme stunt to keep them out of trouble.

Even something as small as a little kitten can spell big trouble for a hognose snake. So the first part of its disguise is to act tough. But if that doesn't work, the snake puts on a show that would impress any Hollywood stuntman. It fakes its own death! And the snake doesn't just play dead, it smells dead, too. It emits such a foul stench that even flies are fooled!

When threatened (top), the hognose snake acts tough (center). If that doesn't work, it plays dead and gives off a terrible stink.

A stuntman stands in for an actor during a dangerous scene (above), while the actor hides behind a rock, waiting for his cue (inset).

Hollywood stuntmen are famous for fooling people. This actor in a western film tells how moviemakers trick the audience:

> *We just performed a scene for a movie in which my horse was shot out from under me. Years ago in pictures, horses were tripped to make them fall, but not today. Hal Needham, who trained this horse, was actually riding it. The horse responded to the pull of a reign and fell on cue. Instead of riding the falling horse, I was waiting safely behind a big rock. When Hal Needham rolled out of sight of the camera, I simply came out shooting. We were dressed alike, so the audience is fooled.*

After having fooled its enemy, the hognose snake gets to return to life. But in the past, some humans weren't as lucky. Until relatively recently, doctors lacked the equipment to be 100 percent sure that their patient had passed away. In the 19th century, premature burial was of such grave concern that alarms were fitted into coffins to be rung by the nearly departed. The fear of premature burial continues today. A coffin alarm patent was issued as recently as 1984. However, the chances of anyone actually using the alarm are remarkably small, thanks to modern embalming techniques.

In the 1800s, people were so afraid of being buried alive that coffins featured a bell that could be rung from the inside.

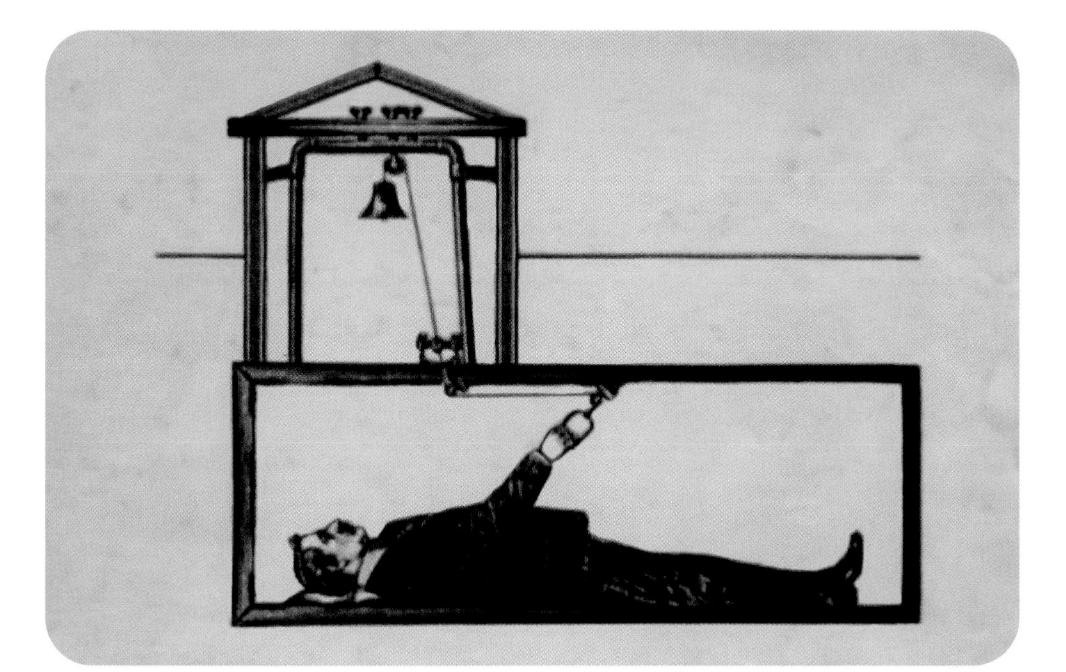

The Caterpillar

If the hognose snake is the ultimate stuntman, our next contender has the costume department all sewn up. With the arrival of spring in the forest, dedicated followers of fashion know that it's time to change into the new season's styles. That's why the caterpillar crawls into number four in the countdown. If it doesn't keep up with the latest woodland style, it'll become a real fashion victim and be eaten!

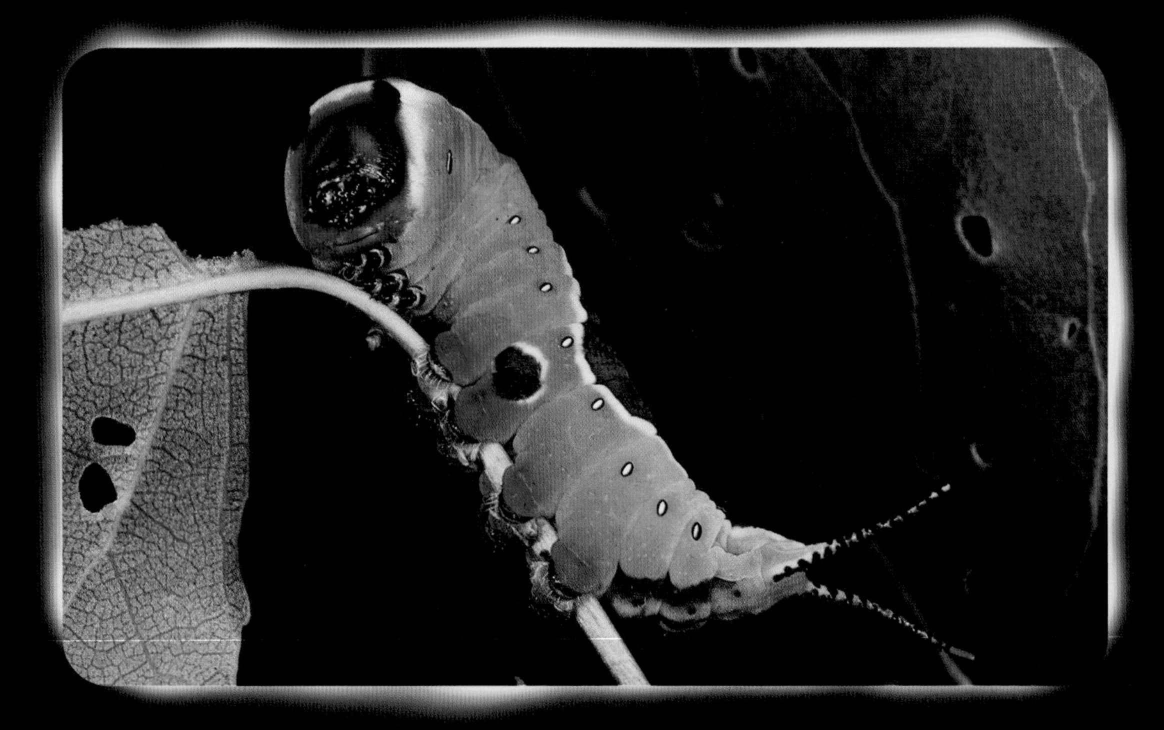

Luckily, the caterpillar keeps its wardrobe up to date by casting off its skin at least four times. Each molt gives it more room to grow and lets it color coordinate with its environment.

Caterpillars are number four in the countdown because their extreme fashion sense leads to some extraordinary disguises. Some caterpillars go for the grunge look and disguise themselves as bird droppings! The idea is to make itself look anything but appetizing.

Another caterpillar is dressed to kill. The South American snakehead caterpillar disguises itself as a snake, complete with antennae that flicker in and out like a forked tongue. For caterpillars, disguise is all about survival. Some humans in disguise are called spies, and sometimes they disguise even their weapons.

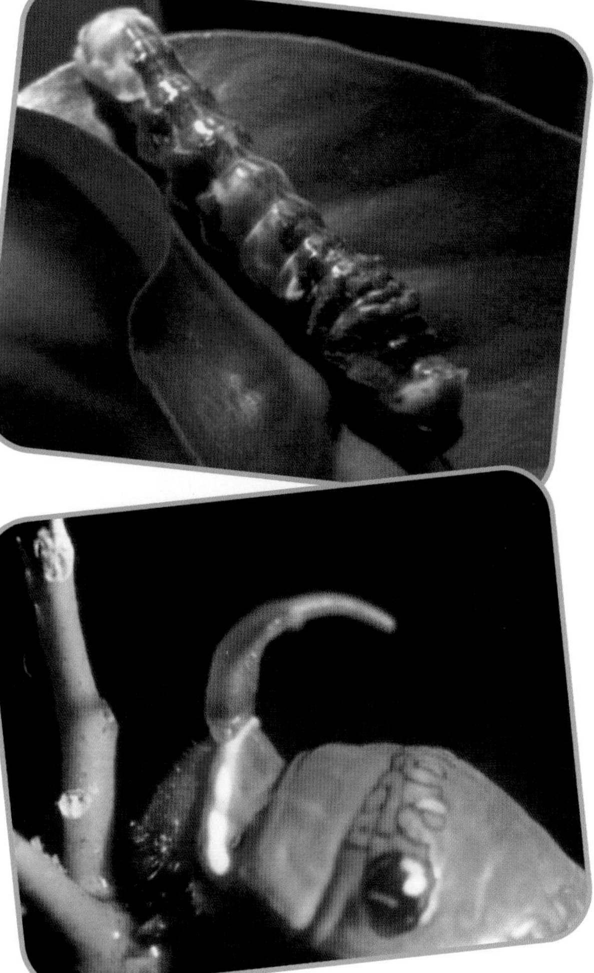

Some caterpillars pose as bird droppings to hide (top), while the snakehead caterpillar of South America disguises itself as a snake (bottom).

The International Spy Museum has all kinds of cool gadgets, including a gun disguised as a glove (right) or as a cigarette lighter (inset).

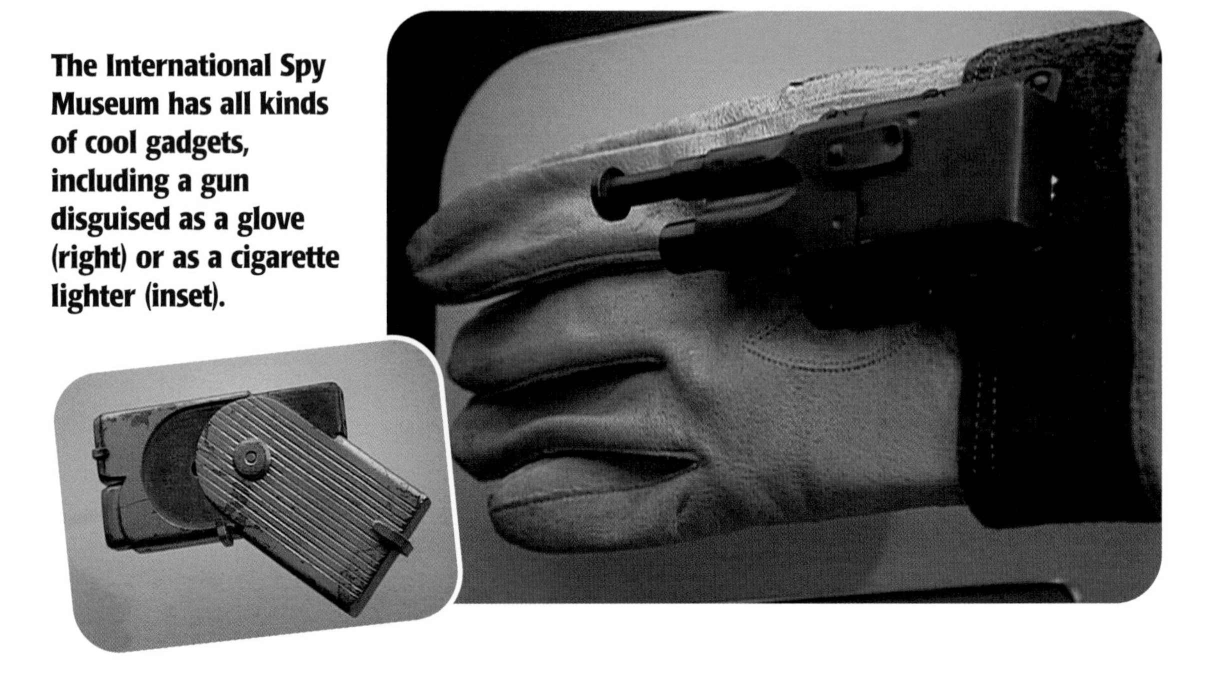

Espionage is a shadowy world full of deception, but if you visit the International Spy Museum in Washington, D.C., you can learn some tricks of the trade. Peter Ernest is the museum director and has some cleverly concealed Cold War weapons on display. You can see a gun disguised as a glove, a cigarette case, a cigarette lighter, and even a flashlight! He describes one such weapon:

> *Most of those devices are defensive in nature. If you had the lipstick gun right now and you tried to shoot me, I don't think your chances would be very good. Frankly, it's very hard to take an object that small and hit much of anything beyond three or four yards away. It did, however, at least give the agent that carried it some sort of defensive device so that if they were in a tight spot they might have a chance of escape.*

Caterpillars may not carry concealed weapons like human spies, but they do have one final trick that would make even James Bond jealous. Caterpillars can change not only their clothes but also their bodies, thanks to the wonders of metamorphosis.

As it creates its cocoon, the caterpillar erases its old identity. Caterpillars are number four in the countdown because their whole life cycle is dedicated to disguise. Even the cocoon looks like a leaf. After two weeks of hiding, the caterpillar finally has a chance to leave its shadowy past behind forever.

Any spy would admire the ability of caterpillars to completely change their bodies as they turn into a butterfly or moth.

3

The Sea Dragon

The next contender in our countdown of extreme disguises could have come straight from the nightmares of the earliest European explorers. Back then, the sea was thought to be full of strange creatures, like the terrifying dragons that appeared on the maps and charts of the time. Believe it or not, a sea monster still lives off the coast of Australia. It's so well disguised that we seldom see it.

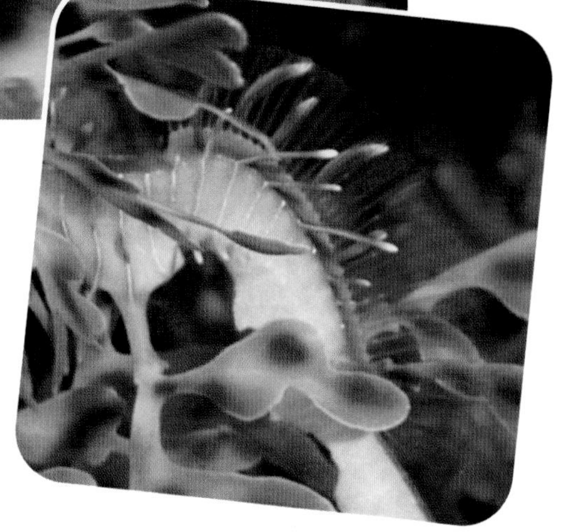

The leafy sea dragon (above) uses its plantlike limbs to fool other sea creatures (inset) .

Coming in at number three in the countdown is the leafy sea dragon. Its camouflage is so convincing that it even fools fish! The sea dragon's body is designed for deception. Those long, leafy limbs have no other function apart from camouflage. Unlike the dragons of ancient mythology, however, this strange fish is no monster. It's actually a tiny cousin of the sea horse and is a threat only to small crustaceans.

Powered by transparent fins, it stealthily glides through the water. It uses its snout like a vacuum cleaner, feeding on animals that are literally sucked in by the dragon's leafy costume.

Instead of the female, it is the male sea dragon that carries fertilized eggs on his tail.

As Kristy Forsgren from the Aquarium of the Pacific in Long Beach, California, explains, the sea dragon's most extreme deception involves some really unusual behavior:

> *The sea dragons also have an additional trick of disguise. It's the males who become pregnant with the embryos! When the male sea dragon is ready to accept eggs, he puts on his disguise. His tail becomes red and swollen, which makes him more noticeable to other animals in the ocean, but also to his mate. When the eggs are embedded onto the underside of his tail, algae will grow on top of the eggs, further camouflaging him while he is pregnant.*

The sea dragon isn't the only one going undercover as a woman to protect the innocent. Some men dressed as females in the name of the law. They were members of the New York Police Department's very own dragnet squad, which kept city streets safe in the 1950s. By masquerading as women, the officers fooled an average of five criminals each day.

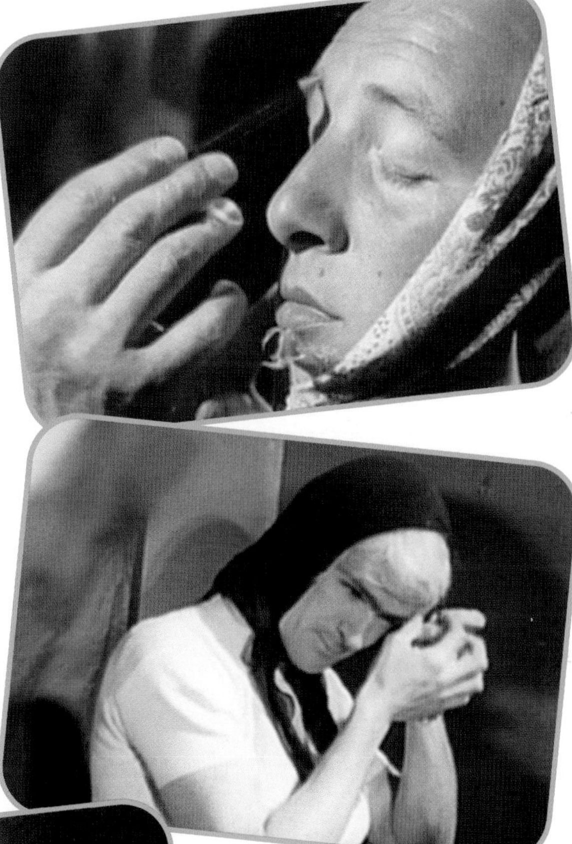

To snag criminals, undercover officers of the New York Police Department dressed as women in the 1950s.

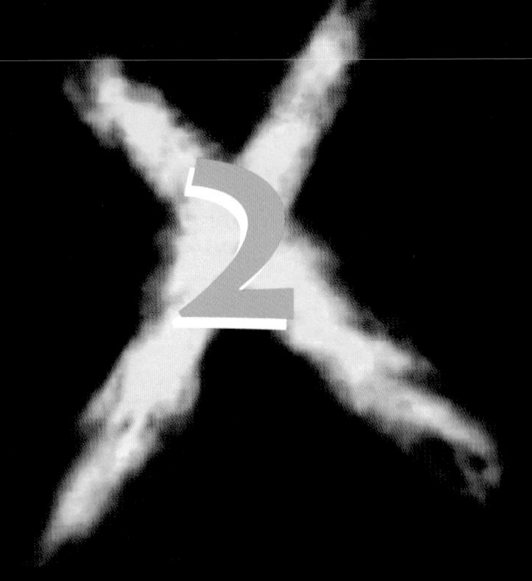

The Walking Stick

Sometimes the best way to stay out of trouble is to keep very, very still. To see the ultimate wooden performance, you just have to visit the Los Angeles Natural History Museum Insect Zoo. Curator Kelli Walker describes a walking stick:

What I have here is a Vietnamese walking stick, one of the 1,000 different species of walking sticks in the world. Walking sticks are nocturnal, so during the day they keep very still. They try to look like a branch on a tree, so they sit on the bark and barely move at all in order to protect themselves from birds that might want to eat them.

Baby walking sticks avoid predators by mimicking ants (top). Adults keep very still to look like a tree branch (bottom).

These insects are number two in the countdown not only because they look like sticks, but also because they once were disguised as ants! Since it takes three months to become giant and prickly, the juvenile has a different disguise. Ants are notorious for attacking other insects, so the baby walking stick takes advantage of that bad reputation by mimicking the way an ant looks and moves. This keeps predators away.

The transformation from baby to adult is extreme. With its twiggy tail and leafy legs, the adult giant prickly walking stick begins its new life as a dead branch.

Back in the 1950s, one man managed to transform his life even more than a walking stick. Ferdinand Waldo Demara was an impostor who pretended to be so many people that his life story was turned into a movie titled *The Great Impostor.*

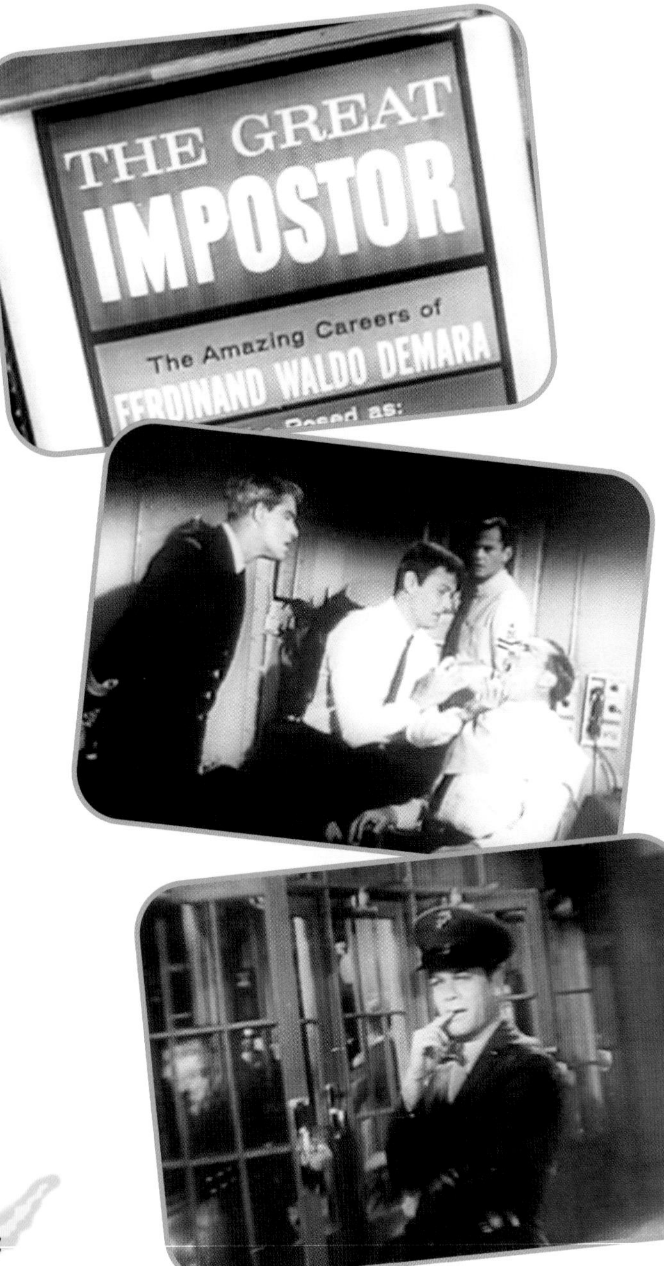

Ferdinand Waldo Demara faked so many identities that Hollywood made a movie of his life called *The Great Impostor.*

Is it an insect or a tree branch? The walking stick perfectly blends in with its surroundings.

The real Demara was unmasked several times, but in the movie, the great impostor just kept moving on to the next impersonation to satisfy his desire to become a hero. For the walking stick, however, a good disguise does much more than just feed its ego. It helps it stay alive!

1

The Octopus

To find the animal that's number one in our countdown of extreme disguises, you have to search the bottom of the sea. These tentacles belong to the ultimate animal transformer: the octopus. It's number one in the countdown because it can change shape, color, and skin texture in less than a second, which is really helpful when there's a hungry shark around.

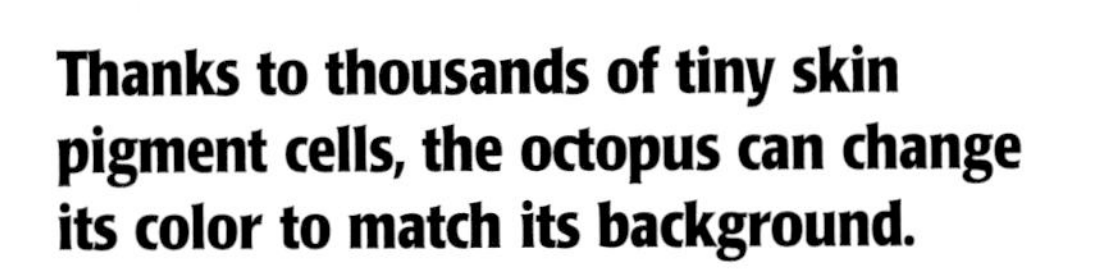

Thanks to thousands of tiny skin pigment cells, the octopus can change its color to match its background.

This remarkable disguise is thanks to thousands of tiny pigment cells in the octopus's skin. By expanding and contracting the cells, the octopus can create the necessary colors for complete camouflage.

Once the Head of Disguise for the CIA, Antonio Mendez made himself look different with each identity he assumed.

Humans can't create an inky smokescreen like the octopus can, but professional disguise artists have other tricks that up until recently have been classified as top secret. Antonio Mendez was the CIA's head of disguise, and he is the first agent to be given permission to share some of the tricks of his trade:

> *If you want to do espionage, you don't do it from an armchair in New York or Washington, D.C. You do it in the field, and in order to do it you have to move people invisibly. You do it with disguise. You change [spies'] identity and have them blend in so they're not seen.*

The difference between real spies and those you see in motion pictures is that the real ones need to be uninteresting, even ugly people, so others won't notice them. The first time I did it in the field was in Karachi, Pakistan. I went down to the local bazaar and bought all the clothing you would wear if you were Pakistani. I put the clothes on, then rolled around on the ground to soil them. I had to make sure I was scruffy enough. When I first went out on the street I had a rush of adrenaline. You get a sense of power because you're invisible, and that's an amazing feeling.

A woman is transformed into a homeless man. Unlike movie spies, real spies try to avoid attention.

A mimic octopus (below) moves just like a real sea snake (right), as another poses as a crab (bottom).

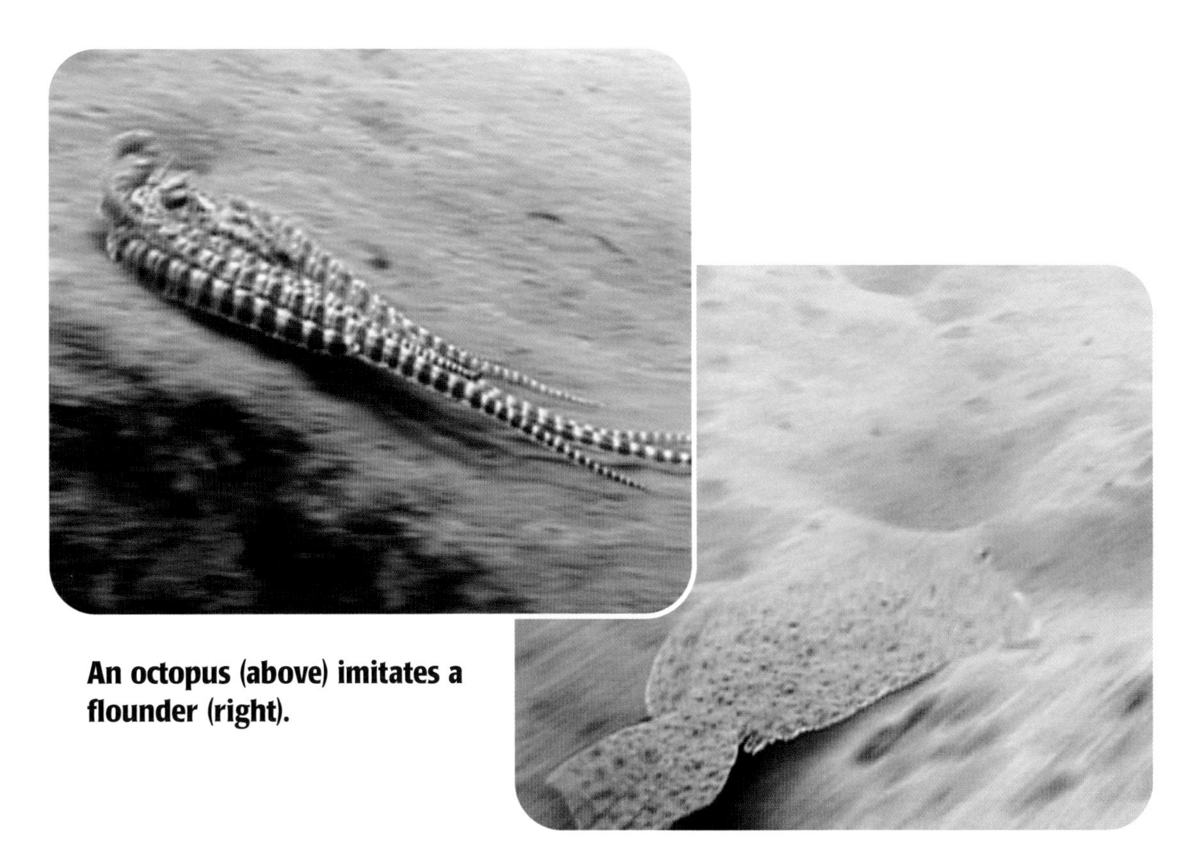

An octopus (above) imitates a flounder (right).

There is one octopus in particular that takes the cloak of invisibility to the extreme: the mimic octopus. It's number one in the count-down because not only can it use disguise to look the part, but it also seems to act the part as well. This has marine biologists asking questions. Is the mimic octopus really trying to swim like a flounder or mimic the shape and color of a sea snake or a deadly lionfish? No one knows how or why this octopus is capable of such extreme mimicry. All we can tell for certain is that when it comes to disguise, the mimic octopus really is The Most Extreme.

For More Information

Bryan and Cherry Alexander, *Journey into the Arctic.* Oxford: Oxford University Press, 2003.

Thornton W. Burgess, *The Burgess Bird Book for Children*. Mineola, NY: Dover, 2003.

Anna Clairborne, *Octopuses*. Chicago: Raintree, 2004.

Kris Hirschman, *The Octopus*. San Diego, CA: KidHaven Press, 2003.

Alice B. McGinty, *Crab Spider.* New York: Rosen, 2003.

Farley Mowat, *Never Cry Wolf: The Amazing True Story of Life Among Arctic Wolves*. Minneapolis, MN: Sagebrush Education Resources, 2001.

David M. Nieves, *Reptiles Up Close.* Kansas City, MO: Reptile Education & Research, 1999.

Ruth Soffer, *Mimicry and Camouflage in Nature*. Mineola, NY: Dover, 2002.

Glossary

adrenaline: a hormone that generates energy

biologists: scientists who study living things

camouflage: a technique or material used to hide from enemies

crustacean: a type of aquatic animal with an exoskeleton such as lobsters and shrimp

espionage: government spying on other governments

feline: related to the cat family

hormonal: related to hormones, substances in the body that produce specific effects

mimicry: imitating others

molt: to shed hair, fur, or other outer layer periodically

pigment: a substance that produces colors

predator: animal that hunts other animals for food

toxins: substances that can cause injury or death

venomous: an animal or insect capable of producing poisonous venom

Index